Back and Forth

By Patricia J. Murphy

Consultants
Martha Walsh, Reading Specialist

Jan Jenner, Ph.D.

Children's Press®
A Division of Scholastic Inc.
New York   Toronto   London   Auckland   Sydney
Mexico City   New Delhi   Hong Kong
Danbury, Connecticut

Designer: Herman Adler Design
Photo Researcher: Caroline Anderson
The photo on the cover shows children swinging back and forth.

Library of Congress Cataloging-in-Publication Data

Murphy, Patricia J.
  Back and forth / by Patricia J. Murphy.
    p. cm. — (Rookie read-about science)
  Includes index.
  Summary: A simple introduction to back and forth movement.
  ISBN 0-516-22552-9 (lib. bdg.)   0-516-26865-1 (pbk.)
    1. Motion—Juvenile literature. 2. Oscillations—Juvenile literature.
  [1. Motion.] I.Title. II. Series.
  QC133.5 .M86   2002
  531'.32—dc21

                                                    2001002687

Move backward. Move forward. You have just moved back and forth.

Do you know how many
things move back and forth?

Mothers rock their
babies back and forth
to comfort them.

Windshield wipers swish
back and forth to clear rain
and snow from car windows.

Bakers spread frosting back
and forth on birthday cakes.

Lumberjacks move saws back and forth to cut logs.

Your arms and legs move back and forth when you make snow angels.

Flags wave back and forth
on flagpoles.

How do things move
back and forth?

Forces such as pushes
and pulls start and stop
moving objects.

You create a force of push when you kick a soccer ball.

When a friend kicks the soccer ball back to you, you are both moving the soccer ball back and forth.

16

Strong winds make trees sway back and forth.

Waves bring shells and sand back and forth from the water to the beach.

19

20

Crickets rub their wings back and forth to make their musical sounds.

Dogs wag their tails back and forth when they are happy.

Objects that *swing* back and forth are called pendulums (PEN-juh-luhms).

Some clocks use pendulums to keep time.

A pendulum needs a push
or a pull to start moving.

First, it swings forward.
Next, it swings back just
as far. Then, it swings
back and forth until
something makes it stop.

Move backward.

Move forward. Keep moving until you stop!

Words You Know

back

forth

lumberjack

pendulum

push

windshield wipers

Index

About the Author

Patricia J. Murphy lives in Northbrook, IL, where she writes children's books. She also writes for magazines, corporations, and museums. She loves to rock her newborn niece, Olivia, back and forth until she falls asleep.

Photo Credits

Photographs © 2002: Corbis-Bettmann: 10 (Richard Hamilton), 12, 31 bottom left (David Martinez), 19 (David Muench), 5 (Karl Weatherly); Peter Arnold Inc.: 23 (Gerard Lacz), 27, 31 top right (Leonard Lessin); PhotoEdit: 24 (Tony Freeman), 3 (Bonnie Kamin), cover (Michael Newman); Stone: 16 (John Callahan), 7, 31 bottom right (Philip Condit II), 11 (Hans Peter Merten); Visuals Unlimited: 6 (Gary W. Carter), 20 (Mary Cummins), 9, 31 top left (Preston J. Garrison), 8 (Mark E. Gibson), 28, 29, 30 top, 30 bottom (RDF), 15 (Inga Spence).